MINING IN CORNWALL
VOLUME FIVE:
THE NORTH COAST

MINING IN CORNWALL
VOLUME FIVE:
THE NORTH COAST

L.J. Bullen

In memory of my late father, Humphrey K. Bullen (1886-1951), who was my mentor and an invaluable source of information. The relatively recent term 'Industrial Archaeology' was unknown to him and his peers – they were genesis.

TEMPUS

Acknowledgements

The preparation of this book has been much assisted by the following persons to whom I would like to record my appreciation and thanks:

Nathan Andrew, Arthur Bosanko, Bruce Grant, Jenifer Harris, Kathy & Anne Miles, Cindy Pearse, Eric Rabjohns, Joyce Tregonning, Colin and Mark Wills.

I owe a particular debt of gratitude to my daughter, Anne Smith, and my son-in-law, Robert Smith, for their sustained and valuable help.

Once again thanks are due to my Publishers who have rendered every encouragement.

First published 2002
Copyright © L.J. Bullen, 2002

Tempus Publishing Limited
The Mill, Brimscombe Port,
Stroud, Gloucestershire, GL5 2QG
www.tempus-publishing.com

ISBN 0 7524 2750 4

TYPESETTING AND ORIGINATION BY
Tempus Publishing Limited
PRINTED IN GREAT BRITAIN BY
Midway Colour Print, Wiltshire

Contents

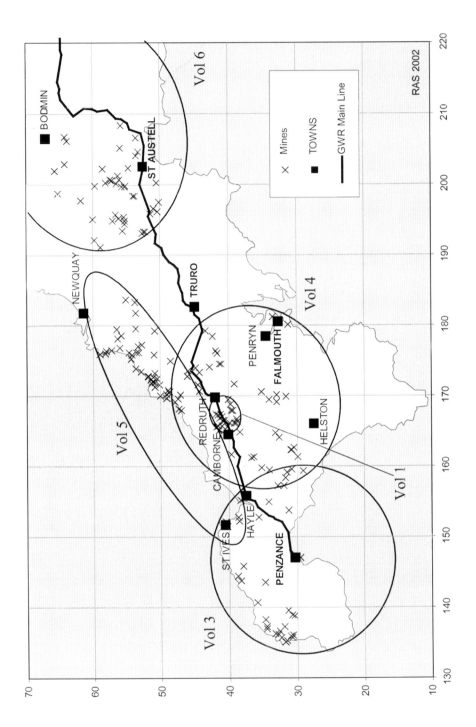

A guide to mine locations referred to in the Mining in Cornwall series (Volume 2 embraces all of the county). Axes show British National Grid References.

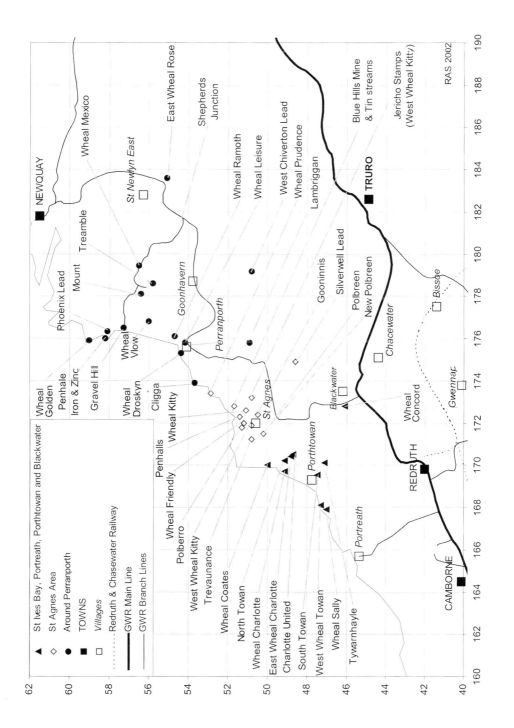

A guide to mine locations referred to in the text. Axes show British National Grid References.

Introduction

This, the fifth book in the series, relates to some of the mines and allied industries in the area between St Ives and Cubert. It is a most interesting and ancient mining district with production covering a suite of minerals. Many of the mines have no recorded history or figures of output. A random compilation of the minerals produced from only eleven mines gives the following figures: Tin 29,720 tons, Copper 153,000 tons, Lead 98,545 tons, Silver 1,481,463 oz, Zinc 102,520 tons, Iron 200,000 tons. It should be stressed that this represents a fraction of the total recorded production of the many mines in the district. A variety of other minerals were mined.

I have included cameos of three mines – Polberro, Wheal Kitty and Cligga, in response to many requests from those who are aware of the range of photographs and supporting archival material in the collection.

The earliest method employed in Cornwall for the winning of tin was by working the valley floors to recover the alluvial material eroded by the agents of weathering. With the commencement of underground mining the dressing floors of the large number of mines deposited their tailings into the streams and rivers. These combined discharges contained sufficient quantities of cassiterite (tin oxide) to make it possible for the tin streaming fraternity to thrive. In recognition of and as a tribute to the valuable contribution which tin streaming played in the story of that metal, the present-day Blue Hills Tin Streams at St Agnes is included in this volume. It represents an accurate portrayal of that ancient and highly skilled vocation. Tin dressing has for long been considered as much an art as a science.

It is felt that the part of the county covered in this volume will particularly appeal to those with an interest in the history of mining. The scenery is impressive and the 'old men's' cliff workings have to be seen to be appreciated. The outcrop of the Perran Iron Lode in the cliffs near the Northern end of Perran Sands is spectacular.

L.J. Bullen, Camborne, Cornwall. February 2002.

Opposite: East Providence Mine, Carbis Bay, St Ives. A view taken in the early 1880s of this unusual engine house with a gable end on the bob wall. It had contained a 40in pumping engine prior to the mine ceasing work in 1881. Note in the background the extensive burrows of the Providence Mine and one of the engine houses.

One
St Ives Bay, Portreath, Porthtowan & Blackwater

Hayle Electricity Generating Station, c.1910. This was situated by the harbour. With the increasing demand for electricity the Generating Station at Carn Brea, Redruth, initially built to supply the Camborne-Redruth Tramway system and a few local mines, was becoming inadequate. The new station at Hayle was built by tidewater and had sufficient surrounding space to expand, as it did in later years.

Above: Hayle Electricity Generating Station 1934. An aerial view showing the river estuary. The vessel at the wharf is discharging coal for use in the boilers of the station. By this time Hayle was connected to the National Grid and the terminal tower of the 132,000 volt grid line can just be seen to the right of the compound on the right hand side of the photograph. (Photo Daphne Pearson)

Opposite bottom: Hayle Generating Station, Cornwall Electric Power Co., *c.*1927. The late A.J. Peller, at that time aged about forty, served his engineering apprenticeship with the Tuckingmill Foundry Co. near Camborne. In 1910 he joined the staff at Hayle, eventually becoming the Deputy Power Station Superintendent.

Above: Cruquius, Holland. The fame of the Cornish pumping engine spread world-wide and in the late 1840s three massive engines were supplied to the order of the Dutch Government. These were to be used to un-water the Haarlem Mere. Two of the engines were supplied by Harvey & Co. of Hayle and the other by the Perran Foundry at Perranarworthal. Fortunately the Cruquius engine has survived, the other two having been scrapped long ago. It is now preserved as a National Monument. These engines had annular cylinders the outer diameter being 144in (3.66m). This photograph shows a general view of the circular engine house and is thought to date from the 1930s.

Left: Cruquius, Holland. An internal view showing the cylinder cover.

Cruquius, Holland. A further internal view showing a part of the cylinder cover and the 'indoor' end of two of the lattice beams.

Cruquius, Holland. Two of the lattice beams protruding from the circular engine house. This group of photographs are a good example of the export of Cornish engineering skills and expertise. These great foundries owed their origins to the indigenous mining i ndustry.

13

Portreath. This north coast port was an extremely busy place which was almost entirely due to its proximity to the inland mines. In this late nineteenth century view it is of particular interest to note the tin stream works and its stack on the beach to the right of the photograph. It was still economic to treat the river for its tin content at this point just before it entered the Atlantic Ocean.

Portreath. A view from lighthouse hill showing the coaster *Islesman* leaving harbour in 1948. Long gone were the great days of the port – coal and building materials being the principal incoming cargoes dealt with at this time and the vessels left in ballast.

Portreath, 1850s. Allowing that this is a poor print taken from Tregea Hill on the western side of Portreath cove it is interesting in that it shows about ten sailing vessels in the port. The harbour was served at this time by a horse-drawn plateway which ran from here to Poldice near St Day. In addition there was a standard gauge (4ft $8\frac{1}{2}$in) steam railway to the Camborne/Redruth mines. Although a relatively small port it was well equipped with excellent storage facilities and rail-borne steam operated cranes.

Portreath, c.1900. A view from the western side of the valley slightly inland of the port. The harbour area extends between the two roads. Careful examination will reveal railway trucks and the steam cranes which operated on the extensive railway network serving the port. Horses were used for shunting operations.

Wheal Sally, near Porthtowan, c.1923. Showing the headgear and steam winding engine house being erected at the commencement of the sinking of a new inclined shaft.

Wheal Sally, near Porthtowan, c.1923. A later view of the incline shaft when sinking was taking place. The waste dump is on the right.

Wheal Sally, near Porthtown, c.1923. The headgear and winder house during the sinking of the inclined shaft. It was always said in mining circles, by persons now long deceased, that the reason for the sinking of this shaft by the Wheal Sally company was never quite understood. It was suggested that there was a possible connection between this and the re-opening of the neighbouring West Towan mine. The promoter of the West Towan scheme also had an interest in Wheal Sally. If there was some financial intrigue the mystery will never be solved this late in time!

Porthtowan, c.1900. The ruinous stack is all that remains at the site of the Engine (or Towan) shaft of South Towan Mine. The mine was a prolific copper producer in the early to middle years of the nineteenth century. It is related that after its closure the engine house on this shaft was blasted down for the stone in order to build a new chapel. One resident of Porthtowan who was more interested in mining than religion commented that he thought it was sacrilege!

Porthtowan, c.1900. Showing a tin streamer working the beach sand for its tin content. The 'White House', as it was named, carries the figurehead of the *Rose of Devon* at the base of its tower. The vessel was a barque wrecked in November 1897 with the loss of all ten crew. The chimney of Ward's Tea Rooms is smoking and on the right is a miner's shay.

Porthtowan, c.1900. The Tywarnhayle valley which runs inland from the porth. A derelict battery of waterwheel-driven Cornish stamps. On the top left of the picture is a part of the engine house on Johns shaft of the Tywarnhayle mine. This had formerly contained a 70in pumping engine. It was moved from here to Wheal Uny, Redruth, in 1874.

Wheal Charlotte. This is situated on the cliffs between Porthtowan and Chapel Porth. The remains of the bob wall of the pumping engine house on Engine shaft in 1978. (Photo Eric Rabjohns)

Wheal Charlotte. On the cliffs between Porthtowan and Chapel Porth. A further view of the engine house on Engine shaft. In the background are the engine houses at Wheal Coates. 1978. (Photo Eric Rabjohns)

North Towan Mine, 1920s. The pumping engine house and, in the background, the engine houses, stacks and extensive dumps of Charlotte United Mine. It is said that around 1820 when Charlotte United was first started it ran out of funds and was about to be prematurely abandoned. The manager's wife, who it appears was a lady with her own private means, offered to put up the cash for one month's further shaft sinking. In this period the mine cut rich and became a successful copper producer for some thirty-five years! (Photo H.G. Ordish)

Opposite bottom: Wheal Coates, Towanroath shaft, St Agnes, *c.*1911. The Sykes contract plant which was used to un-water the mine prior to the installation of the permanent pumping engine. Operations at the mine between 1911 and 1913 produced about eighteen tons of black tin.

East Wheal Charlotte. Situated on the north side of Chapel Coombe, near Mingoose hamlet. c.1908. The mine was re-opened at this time albeit on a limited scale. This photograph was taken from the south side of the coombe and shows the mill site with a set of steam- driven Cornish stamps with buddles below. The stamps are fed with ore from two sources. The rough road at the top allows selected dump material to be brought by horses and carts and discharged into the chute. The tramway from the mine is at a lower level – note the miner pushing a tram on the right. There is an elegantly dressed Edwardian lady in the group near the top of the picture.

Wheal Concord, Blackwater, near Truro, October 1980. The erection of the wooden headgear on Engine shaft. This structure when new had first been erected in 1962 at Contact shaft of Cligga mine, Perranporth. It was then removed to Engine shaft at Nangiles mine in the Carnon valley and finally to Wheal Concord. Nearest the camera are the late Jack and Jill Trounson who were brother and sister. The figure in the group with hands on hips is the late Jack Symons who was the manager of the mine. Nick Worrall, the mine owner, is on the right.

Wheal Concord, Blackwater. The headgear for Engine shaft being lifted on to the concrete bearers already prepared for it in the foreground. October 1980.

Wheal Concord, Blackwater. The new headgear on Engine shaft in the early stages of the re-opening of the mine. The late J.H. Trounson is on the left. He rendered enormous help to the promoters of this venture based on his life-long connection with the Cornish mining industry. 1981.

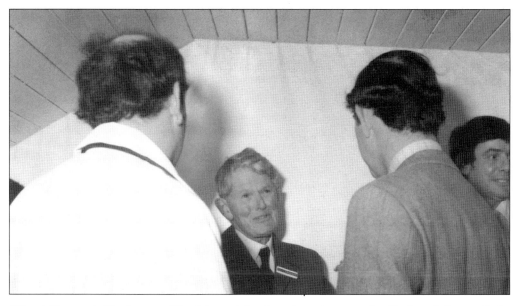

Wheal Concord, Blackwater. The Mineral Lord at the mine was the Duchy of Cornwall. In December 1981 HRH Prince Charles, Duke of Cornwall, paid a visit to Wheal Concord and is seen here speaking with the late Jack Trounson who was at that time Chairman of the Cornish Mining Development Association.

Wheal Concord, Blackwater *c*.1982. All the employees of the mine gathered by Engine shaft.

Opposite: Polbreen Mine, west of the village of St Agnes. In the background is the pumping engine house on New Polbreen Mine which was a complete failure – the lode structures being valueless. However Polbreen, with New Kitty, was a moderate producer of black tin (530 tons) and 250 tons of 6% copper ore. 1930s.

Two
St Agnes Area

Trevaunance Mine, St Agnes, 1976. A close-up view of the totally enclosed beam winder house. Although it tended to be looked upon as a part of the much larger Polberro mine, this mine produced as Trevaunance Consols (1843-1845) 1,180 tons of 4.5% copper ore and as Trevaunance United (1844-87) 550 tons of black tin and 800 tons of 7.5% copper ore. (Photo Eric Rabjohns)

Trevaunance Mine, St Agnes, 1976. A view across the extensive burrows showing the house which had at one time contained a totally enclosed beam whim (winding engine). The mine is generally included in the once very prosperous Polberro mine. Since the date of this photograph the house has been demolished. (Photo Eric Rabjohns)

Gooninnis Mine, St Agnes, 1910. Showing the bob of the 50in Cornish pumping engine on a waggon preparatory to being hauled to the works of the Goonvean & Rostowrack China Clay Co. in the St Austell area. Here it was re-erected to pump from the Goonvean China Clay Pit. This work was undertaken by the Lean family , Engineers (of 'Engine Reporter' fame) and the engine worked at Goonvean until 1956 when it was replaced by electric pumps.

Gooninnis Mine, St Agnes, 1910. The 50in cylinder forms the centre piece of this scene. By Cornish engine standards it is a medium sized cylinder when compared with 90in and 100in cylinders. The engine was being removed to the Goonvean China Clay Pit in the St Austell district. Although out of use since 1956 it is still in situ at the time of writing.

A view of St Agnes from the east, c.1908. On the left is the Gooninnis mine 50in Cornish pumping engine. Polbreen mine pumping engine house is to the right followed by New Polbreen and further right is Thomas's shaft of West Kitty mine with the pumping engine stack smoking. This pumping engine was removed to the Carpalla China Clay Works and when it ceased working there in the 1940s was purchased by the Science Museum in London where it is still in store. (Photo E.A. Bragg)

Polberro Mine, St Agnes, Turnavore shaft *c*.1936. Showing early operations by the Wheal Kitty Co. at this shaft after their decision to abandon the Wheal Kitty, Penhalls and Wheal Friendly areas in 1930. Work is in progress in forming a new shaft collar. The man in white is the late Captain Dick Whitford of St Agnes, a well-known mining figure at that time.

Polberro Mine, Turnavore shaft. c.1936. The men standing on the shaft timbers left to right are G. Wills, Capt. Dick Whitford and J.B. Fern, Mine manager.

Polberro Mine, Turnavore shaft, c.1936. Preparations are being made for the erection of a temporary headgear and shears. It will be noted that the stack of the derelict pumping engine house is smoking. A small boiler had been installed in a lean-to building to provide warm water and the engine house was eventually re-roofed to become the miners dry (change house).

Above: Polberro Mine, Turnavore shaft, c.1936. A further view during the shaft refurbishment to adit level prior to the commencement of pumping.

Right: Polberro Mine, Turnavore shaft, c.1936. The sinking headgear and shears have been erected and are in use. The electric hoist is in the building on the right.

Polberro Mine, *c.*1937. At this time the Californian stamps from Wheal Kitty was being re-erected near the Turnavore shaft in anticipation of production. The scene shows early operations on the site.

Polberro Mine, *c.*1937. Showing the buildings on the new mill site being framed up.

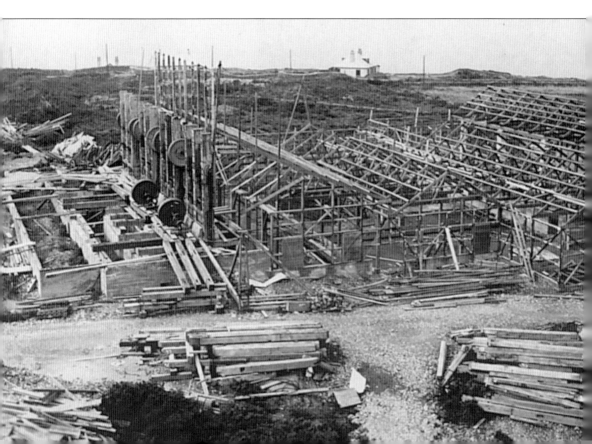

Polberro Mine, *c.*1937. The Mill is being erected and, in the background, the engine house on Turnavore shaft has been re-roofed and has a new lease of life as a dry (miners change house). The permanent headgear has been completed and is in use.

Polberro Mine, *c.*1937. Left to right: the Californian mill building, crusher station, trestle from headgear landing brace to waste dump and Turnavore shaft headgear.

Polberro Mine, *c*.1938. The mill buildings.

Polberro Mine, *c*.1938. Showing the tall building which housed the Californian stamps and beside this the dressing floors. The structure on the right is the crusher station with the inclined conveyor belt which carried the ore to the stamps.

Polberro Mine, Turnavore shaft, c.1937. The permanent headgear is being erected over the temporary structure.

Polberro Mine, Turnavore shaft, c.1937. At this time the shaft was deepened from 644ft to 1,090ft, making it the deepest shaft in the St Agnes district. This is a close-up view of the main legs of the permanent headgear. The temporary sinking headgear and shears are still in use with the electric hoist and capstan in the building on the right.

Polberro Mine, Turnavore shaft, c.1937. The permanent headgear has had its sheave wheel gallows fitted. This facilitates easy wheel changing when necessary. One of the new larger sheave wheels can be seen and to the left of this are two shaft cages brought from Sara's shaft at Wheal Kitty, which is in the background.

Polberro Mine, Turnavore shaft, c.1937. One of the two sheave wheels is in place. Behind the temporary winder/capstan house can be seen two winding drums which have been delivered as part of the permanent winder.

Polberro mine, Turnavore shaft, c.1937. The permanent headgear has its boomstays fitted and both sheave wheels are in place. It will be noted that the temporary winder/capstan house has been shifted to permit the construction of the steel framed building (with an overhead travelling crane) for the permanent winder. The drums of the winder are already in place inside the house.

Polberro mine, Turnavore shaft, c.1938. The engine house now converted to a miners dry has been re-roofed. The permanent headgear and winder are fully commissioned and the crusher station is being constructed in the foreground. Note the base of a jaw crusher at the bottom of the skidway. It would appear that the deepening of the shaft has commenced as indicated by the tram on the trestle to the waste dump.

Polberro mine, Turnavore shaft, c.1938. The waste dump is steadily growing and the crusher station is in the process of being clad.

Polberro mine, Turnavore shaft, c.1939. A view from the top of the Mill building showing the trestle to dump and crusher station which is now complete. The inclined conveyor in the foreground will, when completed, bring the ore from the crusher to the stamps.

Polberro mine, Turnavore shaft, c.1939. Note how the waste dump from the shaft sinking has encroached on the base of the new crusher station and conveyor framework.

Polberro mine, Turnavore shaft, c.1940. Left to right: the Mill building, conveyor to crusher station, blacksmiths shop, workshops, crusher station, headgear, dry and winder house. As a postscript to the foregoing group of photographs of Polberro I would explain that they have been included to show the amount of work which was involved in bringing the mine to the stage it had reached in 1940. The company ran into financial difficulties due to the outbreak of the Second World War. Despite the favourable report of a mining consultant, who estimated that a further £20,000 would permit development to continue, government assistance was not forthcoming. Operations ceased in March 1941. Within twelve months our tin interests in the Far East were lost to the Japanese, thus cutting off one of our principal sources of this valuable metal! I can think of no better conclusion than to quote the late J.H. Trounson who said 'that so much work should have been accomplished and that everything had to be abandoned when success might have been achieved was a great pity'.

Polberro Mine, Turnavore shaft. This photograph was taken many years after the mine was abandoned but the date is uncertain. The perilous state of the shaft timbering is evident. The source of this print is unknown except for the fact that it is rubber stamped on the back Richard H. Bird.

Polberro Mine, St Agnes c.1900. Old men's workings exposed in the cliffs.

Wheal Kitty, St Agnes, c.1908. The erection of the headgear on Sara's shaft. The ancient Wheal Kitty was a Cost Book company until 1904. In 1907 it became known as Wheal Kitty & Penhalls United which also embraced some adjoining mine setts. At this time Sara's shaft was deepened and re-equipped which included the provision of a 65in pumping engine purchased second-hand from the Tindene mine near St Hilary. In addition a new horizontal geared winder was supplied by Holman Bros of Camborne. In the background is the engine house on Turnavore shaft of Polberro Mine.

Wheal Kitty, c.1900. Holgate's shaft. A general view showing the surface balance box of the pumping engine on the left, the headgear, capstan shears and the engine house containing a 50in Cornish pumping engine. This engine was removed to the Parkandillick China Clay Works in the St Austell area in 1912 where it worked until 1953 and is now preserved in working order. On the right of the picture is a typical hand capstan traditionally used for lifting and lowering the pitwork of the pumping engine in the shaft. This type of manually operated capstan was superseded by steam capstans.

Wheal Kitty, Sara's shaft, 1925. The mine closed in 1919 after working, with only a few short stoppages, since 1838. The plant remained on Sara's shaft and this scene was taken about a year before the mine was re-opened. (Photo H.G. Ordish)

Wheal Kitty, *c.*1900. The Cornish stamps. In the right middle ground can be seen the beam winding engine on Reynolds shaft of West Kitty Mine and, over the launder to the left, the capstan shears and stack of the pumping engine on this shaft.

Wheal Kitty, June 1905. The stamps engine at the mine after the disastrous fire on 4 June. This was considered to be arson but the culprit was never discovered.

Wheal Kitty, June 1905. A further view of the stamps engine after the fire. On the left skyline is the engine house of Wheal Prudence.

Wheal Kitty, June 1905. The remains of the Stamps engine house after the fire with smoke still rising from the bob wall.

Above: Wheal Kitty, June 1905. The Stamps engine. The bob and bearings have been hoisted clear of the bob wall in preparation for the rebuilding of the engine house after the fire.

Opposite top: Wheal Kitty, June 1905. The scene after the fire which destroyed the engine house. Following this serious disruption of production the house was rebuilt and the engine, which had miraculously escaped any undue damage, was put to work again.

Opposite bottom: Wheal Kitty, Sara's shaft, *c.*1920. The geared horizontal steam winder built by Holman Bros of Camborne about ten years earlier for use at this shaft.

Wheal Kitty, Sara's shaft, c.1912. A scene at the shaft collar with the steps leading to the plug door of the pumping engine house. This old engine, originally a 60in, was reputedly built to the design of an engineer from St Agnes by the Perran Foundry Co. in the 1860s. Like many engines it was moved from mine to mine and purchased by the Wheal Kitty Co. from the Tindene mine where it had lain idle for some years. It was re-erected on Sara's shaft in about 1910. The conversion to a 65in cylinder engine was achieved by discarding the steam jacket. In this picture work is in progress on the air pump – the bucket is being removed and the floating cover is on the left. By a strange coincidence the life of this engine ended in the parish in which it was designed.

Right: Wheal Kitty, Sara's shaft, *c*.1916. The gearwork of the 65in engine.

Below: Wheal Kitty, Sara's shaft, *c*.1924. A photograph taken when the mine had been lying idle since 1919. Left to right: the house containing the Holman built horizontal steam winding engine,the headgear, capstan shears and the 65in pumping engine house. In the centre background is the boiler house, the boilers from which steamed both pumping and winding engines. At this time the plant was on care and maintenance.

Wheal Kitty, Sara's shaft, c.1926. Preparations are being made to recommence operations at this site. The original headgear was found to be suffering from rot and was taken down. Unfortunately during the dismantling the capstan shears was struck and came to rest against the front of the engine house as shown here. The replacement headgear was purchased second-hand from South Crofty mine where it had formerly stood on Palmer's shaft. It was too short for the requirements at Sara's shaft and therefore had to be reconstructed but using some of the original timbers. By the time it was completed as seen here it was virtually a new headgear. When the mine closed, this headgear was moved to the Turnavore shaft at Polberro Mine and was finally purchased by the company re-opening the Cligga Mine at Perranporth where it was erected on Contact shaft. A much travelled and altered headgear! On the left-hand margin of the picture the new chimney stack for the two new Lancashire boilers is being erected. At this time it was decided to abandon the previous arrangement of one group of boilers supplying steam for both the pumping and winding engines. This new boiler installation was to provide steam for the winding engine and a new Bellis & Morcom vertical air compressor. Beyond the boom stays of the headgear can be seen the work being carried out in connection with the enlargement of the pumping engine boiler house to provide a miners dry.

Wheal Kitty, Sara's shaft, *c.*1926. A slightly later scene than the last photograph showing the new headgear fully completed with the handrails around the sheave platform. The capstan shears has been restored and secured in its correct position. The new stack is nearing completion and one of the two new Lancashire boilers is in place.

Wheal Kitty. Sara's shaft, *c.*1926. The new stack is complete and the boiler house and compressor house are having their roofs laid. By the compressor house can be seen the air receiver.

Wheal Kitty, Sara's shaft, 1927. The pumping engine is at work and all the surface refurbishment completed. The stack of the winder/compressor boilers is smoking indicating that work is proceeding underground. In the foreground the twenty head Californian mill is under construction having been re-sited from its original position some distance away from the shaft. At a later date the mill was enlarged by the addition of a further twenty head of stamps.

Wheal Kitty, Sara's shaft, *c*.1927. Left to right: the electricity substation (the mill was electrically powered), the stack for the boilers which steamed the winding engine and air compressor, compressor house, (behind which is the Mill building), Blacksmiths shop, headgear, capstan shears and pumping engine house.

Wheal Kitty, Sara's shaft, late 1920s. Taken from the Mill building and showing the crusher station with the inclined tramway from the ore bin to the stamps.

WHEAL KITTY MINE. 3.

Wheal Kitty, Sara's shaft, c.1927. A view from the south, with the Mill building in the background on the left and nearer the camera the crusher station with the wooden trestle carrying the tramway to the landing brace in the headgear. The 65in pumping engine is at work and it is clearly a winter scene as the chimney on one of the shaft top buildings is smoking.

Wheal Kitty, Sara's shaft, c.1927. A view from the north showing the Mill buildings nearing completion. This print is interesting in that it shows, behind the headgear, the elevated tramway, now discarded, which led from the shaft to the former Mill site.

Wheal Kitty, Sara's shaft, late 1920s. In the back row fourth from left is George Wills, grandfather to Colin Wills of the present Blue Hills Tin Streams. Front row right is Billy Wills another member of the family. (Photo courtesy Colin Wills)

Wheal Kitty, Sara's shaft, 1929. Taken from the western side of Trevaunance Coombe. The smoking stack on the left-hand side of the scene serves the calciner flues. Continuing to the right are the dressing floors, Stamps, Crusher station, Winder house, stack for the boilers of the winder/compressor and, behind this, the headgear and pumping engine house.

Wheal Kitty, Sara's shaft, *c.*1938. Taken at the time of the scrapping of the 65in pumping engine. Before removing the engine they availed themselves of the opportunity to recover the pitwork above adit level using blocks attached to the nose of the beam and a lorry running to and fro acting as a substitute 'capstan'. Here we see an 'H' piece which had just been raised to surface. The figure in a trilby hat and an arm in a sling is the late Arthur Jory of South Crofty Mine.

Wheal Kitty, Sara's shaft, c.1938. At the time of the scrapping of the 65in pumping engine with a group looking at a fragment of the broken beam. On the far left is the late Arthur Jory, the well-known pitman foreman of South Crofty Mine. The Redruth firm of S.J. Andrew & Son had purchased the engine for scrap and employed Arthur Jory to superintend the demolition work. Sixth from left is Tony Andrew and eighth from left is Harold Andrew who were brothers. Third from right is Bert Congdon. The bowler hatted figure is the late Joseph Blight, then retired, who had been the Chief Engineer at South Crofty Mine for many years. On the extreme right is the late Jack Trounson, then a young man, and now remembered as a Mining Consultant and doyen of the history of Cornish mining and engineering.

West Wheal Kitty, St Agnes, early 1900s. On the left is Thomas's shaft with its pumping engine house, capstan shears and headgear. On the extreme left can be seen the front of the house of the beam winding engine which hoisted through this shaft. The spire of St Agnes church is right of centre.

West Wheal Kitty, Reynolds shaft, c.1900. Showing the pumping engine, headgear and capstan shears on the left and the beam winding engine on the right.

West Wheal Kitty, Reynolds shaft, early twentieth century. The engine house containing a 50in pumping engine with the top of the capstan shears and headgear just visible behind the dwelling house on the left. This was the principal shaft of the mine for many years.

West Wheal Kitty, Reynolds shaft, early twentieth century. A view from Fore Street, St Agnes, with the spire of the parish church left of centre. Immediately to the left of the pumping engine house can be seen the stack of the winding engine.

West Wheal Kitty, c.1900. A group of miners and three agents (senior officials). Among the hierarchy of the mines the type of hat worn was almost like a badge of office. Consequently the agent on the right of this scene is the most senior of the three as indicated by his head-dress!

West Wheal Kitty, late nineteenth century. A view of the fitting shop. It will be noted that a small horizontal steam engine provided power via line shafting and belts to the various machines in the shop.

West Wheal Kitty, late nineteenth century. A posed scene outside the Blacksmiths shop showing work in progress on a large kibble. Prior to the introduction of skips and cages which ran in guides through the shaft the kibble was the standard method of raising ore from underground.

Above: Jericho Stamps in Trevellas Coombe, St Agnes, early twentieth century. When West Kitty Mine commenced production in about 1860 it was impossible to set up a stamps and dressing floors near the shafts as they were sited virtually in the village of St Agnes. Therefore a decision was made to erect their treatment plant at Jericho in the upper reaches of Trevellas Coombe, which entailed conveying the ore over a mile by road. The first stamps were driven by a waterwheel, but as production increased a rotary Cornish beam engine was installed to drive a larger stamp battery. In this view we see the empty house of the first steam stamps which had by this time been superseded by a new set of Californian stamps. This plant together with more up-to-date dressing floors had been installed in the buildings at the rear. Holman Bros of Camborne provided a horizontal steam engine to power the new mill.

Below: Jericho Stamps, early twentieth century. A closer view of the West Kitty stamps. On the road to the left of the stack there are horses and carts which were the means of transporting the ore between the mine and this stamps site. The exhaust steam from the engine which drove the Californian stamps is in evidence. It would appear that building work and alterations are in hand at the lower dressing floors on the left of the picture. A stack made from old boiler tubes has been erected at the end of a new flue. In the corner of the field on the left is a hayrick – a part of the rural scene which has long since disappeared.

Jericho Stamps, early twentieth century. Showing the three calciners left of centre with their separate flues converging into a single flue leading to the scrubber which is on the right of the stack. The scrubber provided a damp alkaline environment which was achieved by running water over limestone blocks. The gases from the calciner flue passed through the scrubber and were then exhausted via the stack to atmosphere. The calciners were used to roast the final tin concentrate in order to remove the arsenic content. The mine obtained a better price from the smelters for the tin oxide and was able to sell the arsenic as a by-product.

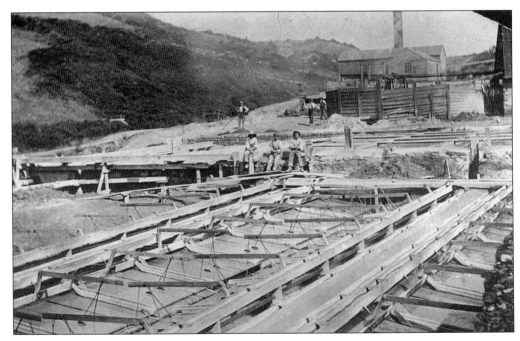

Jericho Stamps, c.1900. A view of the rag frames which recovered the fine tin in the final stage of the dressing floors.

Jericho Stamps, 1930s. A photograph of the Cornish stamps engine house taken many years after the engine had been removed and replaced by Californian stamps. In fact by this time all the other plants had been scrapped following the closure of West Wheal Kitty in 1916. (Photo H.G. Ordish)

Penhalls Mine, St Agnes, late nineteenth century. On the left is the engine shaft. The centre of the picture is occupied by the Cornish stamps and the dressing floors. The house to the right of this contains a winding engine. In the centre background is the Count house. From 1907 the mine was amalgamated with Wheal Kitty.

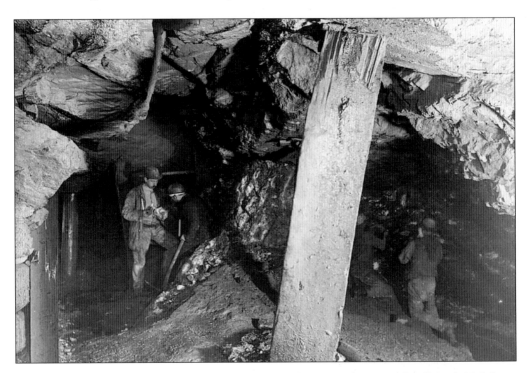

Above: Blue Hills Mine, St Agnes, 1890s. Heave looking west between the 50 and 66 fathom levels. A heave (or fault) causes the lode to be dislocated. (Photo J.C. Burrow FRPS)

Opposite top: Blue Hills Mine, St Agnes, 1890s. Heave looking east. (Photo J.C. Burrow FRPS)

Opposite bottom: Blue Hills Mine, St Agnes, 1890s. Timber men at work in stopes above the 66 fathom level. The lode here is very flat. (Photo J.C. Burrow FRPS)

Blue Hills Mine, St Agnes, 1890s. An underground scene. (Photo J.C. Burrow FRPS)

Blue Hills Mine, St Agnes, 1890s. A group of St Agnes miners captured for posterity in an underground scene well over a hundred years ago. (Photo J.C. Burrow FRPS)

Trevellas Coombe, St Agnes, early 1900s. This shows the first set of tin streams down the valley below the West Wheal Kitty dressing floors at Jericho stamps which is in the background. The rag frames and round tables form the basis of these operations. (Photo E.A. Bragg)

Blue Hills Mine, St Agnes, 1945. A view from the western side of Trevellas Coombe. The engine house once contained a 70in pumping engine. (Photo H.G. Ordish)

Trevellas Coombe, St Agnes, early 1900s. A scene further down the valley showing extensive tin streaming works covering nearly the whole of the valley floor. From the left hand edge of this scene the valley extends for a further half a mile to the sea all of which contained tin streams and yet another waterwheel driven stamps. As a matter of interest the present Blue Hills Tin Streams is centred on the derelict stone wall situated bottom left in this photograph. At that time the wheel had been removed. (Photo courtesy Royal Institution of Cornwall)

Trevellas Coombe, St Agnes, c.1932. A view looking inland and showing the last stamps before the beach. (Photo H.G. Ordish, courtesy Royal Institution of Cornwall)

Trevellas Coombe, St Agnes, c.1932. Taken looking down the valley to the beach showing the last stamps site. (Photo H.G. Ordish, courtesy Royal Institution of Cornwall)

Trevellas Coombe, St Agnes, c.1934. A further view of the last stamps site before reaching the sea which was locally known as Trezise stamps. This photograph shows the two rotary pulverisers in use as part of the dressing process. (Photo H.G. Ordish)

Trevellas Coombe, St Agnes, early 1950s. This is the site of the present Blue Hills Tin Streams and should be compared with later photographs in this volume showing the works as it is today. The figure is the late Jim Green.

Trevellas Coombe, St Agnes, April 1938. The site of the present Blue Hills Tin Streams showing the tin streamer using a typical long-handled Cornish shovel to fill a standard type Cornish wheelbarrow. All the tin streams were individually owned and operated, some by small companies and others by a group of working partners. Their common objective was to recover the tin which had escaped into the river from the dressing floors of the mine. (Photo H.G. Ordish, courtesy Royal Institution of Cornwall)

Trevellas Coombe, St Agnes, 1966. A further scene showing another period, this time of dereliction, at the present Blue Hills Stamps. (Photo Charles Woolf)

Trevellas Coombe, St Agnes, 1966. A view of the overgrown site from the track above. (Photo Charles Woolf)

Trevellas Coombe, St Agnes, 1966. This photograph gives some indication of the enormous task undertaken by the Wills family, the present owners, to rehabilitate the stamps to its present state. (Photo Charles Woolf)

Trevellas Coombe, St Agnes, 1966. The scene of dereliction should be looked at in conjunction with the earlier photographs in this volume showing the valley as a hive of industry.

Blue Hills Tin Streams, Trevellas Coombe, St Agnes, 2001. Contrary to popular belief, tin is still won in Cornwall. Colin Wills and his son Mark run the tin streams at Trevellas which has been refurbished in stages since 1994. In this work they have been ably encouraged and supported by Colin's wife, Hilary. This scene shows the waterwheel purchased from South West Water formerly at Tregurra, Truro, and rebuilt by Mark Wills at Trevellas. (Photo Cindy Pearse)

Blue Hills Tin Streams, St Agnes, 2001. Showing the six head of stamps currently in use. The remaining six are to be restored in 2002. Note the pair of home-made dipper wheels. The right hand wheel is raising the material from the stamps and delivering it by a launder (gutter) to the buddle on the right. The left-hand wheel lifts the slimes from the buddle to the settling tank or lagoon. (Photo Cindy Pearse)

Blue Hills Tin Streams, St Agnes, 2001. Showing the launder bringing the water to the wheel and the tops of some of the six stamps lifters appearing beyond the massive stone wall. (Photo Cindy Pearse)

Blue Hills Tin Streams, St Agnes, 2001. Mark and Colin Wills the father and son team who have achieved a remarkable feat at this site. It is now open to the public all year round. (Photo Cindy Pearse)

Blue Hills Tin Streams, St Agnes, 2001. A view from the back of the stamps showing the feed hole for the six head in use. The extended barrel and cams await the re-erection of the other six head which will return the battery to its full capacity. (Photo Cindy Pearse)

Blue Hills Tin Streams, St Agnes, 2001. A further photograph taken from the back of the stamps showing the stamp lifters and beyond them the barrel and cams. (Photo Cindy Pearse)

Blue Hills Tin Streams, St Agnes, 2001. The stamp heads which, by repeatedly rising and falling on to the rock feed, will reduce it to sand. Running water is essential in the process as a vehicle for the pulp throughout the dressing operations. (Photo Cindy Pearse)

Blue Hills Tin Streams, St Agnes, 2001. Another view of the stamps and dipper wheels already described. (Photo Cindy Pearse)

Blue Hills Tin Streams, St Agnes, 2001. The buddle which is fed at the centre by the wooden launder bringing the water borne material from the stamps. A power take-off from the stamps axle drives the three wooden arms. From each of these a suspended bar, with brushes fixed on the underside, ensures an even distribution of the pulp feed emanating from the centre. When the circular buddle is full the feed is stopped and the drive to the brushes disengaged. (Photo Cindy Pearse)

Blue Hills Tin Streams, St Agnes, 2001. Another view of the buddle. When the circular enclosure is full the tin concentrate will be distributed with the heavier tin ore (i.e. the higher percentage) nearest the centre and reducing in value by stages to the outside of the pit. The experienced tin dresser will then decide where to delineate between the head, middlings and tails after which the pit is dug out accordingly. (Photo Cindy Pearse)

Blue Hills Tin Streams, St Agnes, 2001. A view of the plant excluding the waterwheel which is behind the wall on the left. Although water power is extensively used there is also a great deal of hard manual labour involved in these operations. The faithful reproduction of the traditional Cornish pattern shaped ends to the wooden cap piece over the buddle provides a nice touch. (Photo Cindy Pearse)

Blue Hills Tin Streams, St Agnes, 2001. Mark and Colin Wills are standing in front of a shaking table fitted with a Holman Bros of Camborne patent motion whilst behind them is a ball mill which feeds the table. This is the next process which again produces ore in three categories i.e. heads, middlings and fines (tails). The various stages are continually upgrading the final product. (Photo Cindy Pearse)

Right: Blue Hills Tin Streams, St Agnes, 2001. Colin Wills is by a substitute kieve which is made of metal and came from a smelting works. The usual Cornish kieve was made of wood with iron strapping. He is using a vanning shovel the use of which calls for a high degree of skill. By careful and experienced manipulation a sample of tin ore can be 'thrown' on the face of the shovel into its various grades. This is one of the demonstrations provided for visitors to the tin streams. (Photo Cindy Pearse)

Below: Blue Hills Tin Streams, St Agnes, 2001. The concentrate is smelted on site and this photograph shows the metal as smelted together, with tin ingots, jewellery and other items. A range of artifacts made of tin are available. The tin ore has been refined to a concentrate, smelted to the tin metal and fashioned into an array of products all within the confines of the works. (Photo Cindy Pearse)

Trevaunance, St Agnes, c.1900. The buildings and stack have been used at various times as a foundry, tin stream works and a factory. On the right of the picture is the engine shaft of Penhalls Mine whilst nearly on the skyline to the left of this is a horse whim. On the skyline left of centre is the Wheal Prudence 70in pumping engine which had formerly worked at South Towan mine, Porthtowan. Just to the right of the pumping engine house the winding engine house is visible. Wheal Prudence is now within the boundary of the RAF airfield constructed during the Second World War. On the left skyline across the bay are the engine houses at the Penhale Iron Mine not far from the outcrop in the cliffs of the Perran Iron lode. In the centre left of the photograph can be seen the roof of a house and slightly to the left of this is the mast of a sailing vessel in St Agnes harbour.

Wheal Friendly, St Agnes, early twentieth century. Showing the 60in pumping engine, headgear, landing brace and trestle which carried the tramway to the dump and stamps. Traditionally in Cornwall, the weekly washing was always done on Monday. I think we can say that this photograph was taken on a Monday as careful study will reveal that there are three houses with lines of washing hung out to dry!

Wheal Friendly, St Agnes, early twentieth century. A view from the Eastern side of Trevaunance Coombe showing the considerable waste dump and the tramway trestle to the Mill. This consisted of two head of pneumatic stamps. The shelving killas on which the stamps were constructed could not withstand the heavy blow which this type of stamp created, with the result that the whole stamp battery was sliding down the hillside! Consequently it had to be abandoned and all ore sent by horse and cart to Jericho stamps for treatment. It should be mentioned that Wheal Friendly was working as part of West Wheal Kitty at this time. In the background can be seen the empty engine house on the Turnavore shaft of Polberro Mine where the Wheal Friendly pumping engine originally worked.

83

ST. AGNES BEACH, CORNWALL

Above: Wheal Friendly, St.Agnes, early twentieth century. A view of the mine from the beach. The two boys on the rocks have been for a swim in the Atlantic and are towelling themselves dry – a time honoured custom of Cornish lads. (Photo S.J. Govier)

Opposite top: Holman Bros Ltd, Camborne, early twentieth century. The erecting shop at the No.1 Works of this well-known firm showing a twin cylinder horizontal geared steam winding engine. This was destined for the Wheal Friendly Engine shaft. (Photo J.C. Burrow FRPS)

Opposite bottom: Wheal Friendly, St Agnes, late 1920s. In the 1926 re-opening of Wheal Kitty the intention was to work the mine in conjunction with Wheal Friendly, Polberro and other mines on the western side of Trevaunance Coombe. The un-watering of Wheal Friendly was an early priority in this scheme. At first it was planned to use a Cornish pump – indeed the 70in engine then standing idle on Harvey's Engine shaft at South Tincroft mine was purchased with this in mind. However before very much work had been done there was a change in policy which resulted in the use of an electrical pump instead. Here we see the pump delivered on site. In the background is the pumping engine house on Wheal Friendly engine shaft. The V twin James motorcycle was owned by the late J.H. Trounson, who took the photograph.

Wheal Friendly, St Agnes, late 1920s. Showing the short headgear erected and braced into the bob wall of the former pumping engine house. The two sheave wheels served the ropes of the capstan for lowering the pump and a cage for raising and lowering the men. The drum of the capstan can be seen on the waggon.

Opposite: Cligga Mine, Perranporth, *c.*1939. A scene at the time the mine was re-opened and Contact shaft was brought back into use. These early operations were conducted on a small scale and any ore was treated by using a waterwheel driven stamps in Perran Coombe. Note the windlass on the shaft in the background.

Three
Perranporth & District

Cligga Mine, Perranporth, c.1939. A further view shortly after the re-opening of the mine had commenced. The stacks of the vertical boilers, which supplied steam to the small hoist, are on the left whilst further right is the first headgear on Contact shaft.

Cligga Mine, Perranporth, c.1939. Showing left to right the headgear which hoisted on the incline cut into the cliff face (a photograph of which appears in Volume 2 of this series), the house containing the steam hoist and boilers, contact shaft headgear and, if one looks carefully, the 'lander' can be seen in the headgear. His duty is to empty the skip on its arrival at surface.

Cligga Mine, Perranporth, *c.*1940. By this time the Rhodesian Mines Trust Ltd had taken over the property and commenced erecting a Mill for treatment of ore on the site, which can be seen under construction.

Cligga Mine, Perranporth, *c.*1940. A little later scene showing the headgear on the cliff incline, Contact shaft winder and headgear with the new Mill buildings in an advanced stage of construction.

Cligga Mine, Perranporth, c.1940. A ladder way with sollars constructed through an 'old mens' stope during the early period of the re-opening of the mine.

Cligga Mine, Perranporth, c.1943. Showing the new headgear (ex Polberro Mine) on Contact shaft and the complex of mine buildings.

Cligga Mine, Perranporth. *c.*1944. The employees of the mine against the background of the headgear on Contact shaft. (Photo courtesy Colin Wills)

Cligga Mine, Perranporth, *c.*1946. After the closure of the mine in 1945 the plant was scrapped within a relatively short time. The building from this part of the Mill had already gone but the machinery including dipper wheels remained. The tide is out exposing the long expanse of Perran sands in the background.

Cligga Mine, Perranporth, *c.*1946. The late Jack Trounson is examining the Mill engine at the time when the machinery was being scrapped. This engine had formerly worked at Great Wheal Busy, Chacewater.

Cligga Mine, Perranporth, *c.*1946. Another view of the Mill engine which had previously driven the Mill at Great Wheal Busy, Chacewater in the early twentieth century re-working of that mine. Having lain idle for many years it was purchased in about 1939 for use in the Mill at Cligga. This was a single cylinder engine with drop valves. The figure is the late J.H. Trounson.

Cligga Mine, Perranporth, c.1946. Taken at the time of the scrapping of the Mill plant. On the left is a tandem compound steam engine which had previously worked at the Lambriggan Mine. In 1939 it was purchased and installed at Cligga as a Mill engine. The other Mill engine described in an earlier photograph is on the right.

Cligga Mine, Perranporth, A group of visitors to the mine thought to have been taken in the late 1940s and some years after abandonment. The figure wearing a miners helmet and lamp is Frank D. Woodall a well-known model engineer and historian. On the back of this print is written 'a rise 310 feet in from the portal of the level in the cliff.'

Cligga Mine, Perranporth, 1950s. Taken at a shaft station in Contact shaft. Note the suspended compressed air line on the right. The figures are left to right: Ken Hammill, Joe Bendle, -?-, Grenville Pryor, -?-.

Left: Cligga Mine, Perranporth, 1954. A view taken from just inside the portal of a level driven in the cliffs. Note the scree of waste rock in the background which had been dumped from another level higher up the lofty cliffs. (Photo L. Halls)

Opposite: Cligga Mine, Perranporth, 1950s. A group photograph taken in one of the long abandoned levels. Left to right: Ken Hammill, -?-, Donovan Wilkins, -?-, Grenville Pryor.

Cligga Mine, Perranporth, October 1962. At this time Geevor Tin Mines Ltd of Pendeen took a lease on the Cligga sett with a view to assessing the potential of the property. A new wooden headgear was fabricated at Pendeen and brought to Perranporth where it is seen being erected on Contact shaft. (Photo courtesy Jenifer Harris)

Above: Cligga Mine, Perranporth, October 1962. The new headgear is complete and has temporary wheels at this stage. This work was undertaken by Foraky, Mining Contractors whose sign is in evidence. It is interesting to note that the lady who is the source of these two images is the daughter of the late J.B. (Jimmy) Hooper, the engineer in charge of these operations. (Photo courtesy Jenifer Harris)

Right: Cligga Mine, Perranporth, November 1962. The new headgear on Contact shaft in use with an electric hoist.

Cligga Mine, Perranporth, November 1962. A view of the shaft collar showing the wire rope guides, kibble and jockey (crosshead) from which the kibble is suspended. The shaft doors are closed. Refurbishment of the shaft was taking place at this time.

Cligga Mine, Perranporth, November 1962. Showing the headgear and winder house on Contact shaft. In the background is a Nissen hut which served as a stores. During this last period of work at the mine the shaft was deepened below adit level.

Wheal Vlow, Perranporth, 1927. Situated in the sandhills east of Perranporth, the mine was also known at one period in its history as Perran Consols. In the 1920s Wheal Vlow was one of a number of mines which the Anglo Oriental Corporation re-opened in Cornwall. This scene was recorded very early in the operations centred around Hallett's shaft. In the background are the new offices, workshops and the miners dry. The temporary tramway in the foreground is leading to Hallett's shaft.

Wheal Vlow, Perranporth, 1927. Taken a little later than the previous photograph and showing a number of the buildings completed. Left of centre is a steel-framed building being clad, which is the electricity power station containing Paxman vertical high speed diesel engines which drove the dynamos. To the right of this is a temporary wooden headgear on Hallett's shaft then being refurbished.

Wheal Vlow, Perranporth, 1927. During the un-watering of the mine at this time, the pitwork of the 60in Cornish pumping engine which operated at this shaft in the previous working was hauled out. On the left are old main rods together, with on the right, castings of door pieces and pumps laid out on the sand.

Wheal Vlow, Perranporth, c.1928. This shows a further selection of items removed from Hallett's shaft during the shaft recovery. In the foreground is the barrel of an old windlass (or 'tackle' as the Cornish miner termed it) with the remains of the rope still wound on. To the left is a sinking door piece and a pump which also formed a part of the pitwork of the pumping engine.

Wheal Vlow, Perranporth, *c*.1928. The clearing of Hallett's shaft produces even more relics from the previous working of the mine. In the pile is a small sheave wheel and pitwork items include a sinking bucket, clack seat and old chains.

Wheal Vlow, Perranporth, *c*.1928. The steel headgear on Hallett's shaft is being erected together with the ore bin. The building on the right is the electricity generating station.

Wheal Vlow, Perranporth, c.1928. Hallett's shaft. Left to right: miners dry, electricity generating station, headgear and ore bin being erected.

Wheal Vlow, Perranporth, c.1928. Hallett's shaft. Left to right: generating station, electric winder house, headgear and ore bin.

Wheal Vlow , Perranporth, c.1928. The surface plant of the mine is nearing completion. Left to right: winder house, headgear and power station. Note the water cooling plant at the back of the power station for the diesel engines. This re-working of the mine did not result in any production.

Wheal Vlow, Perranporth, c.1928. The mine is now fully equipped and this photograph shows, left to right: barracks for the miners, electricity power station, headgear on Hallett's shaft, miners dry, workshops etc. It should be explained that the barracks were built to accommodate men who lived some distance from the mine and elected to return home only at week-ends. In many cases this was a necessity primarily because of the time spent in travelling and cost. In the background is the vast area of sandhills between Perranporth and Cubert.

Wheal Vlow, Perranporth, *c.*1929. An internal view of the electricity generating station showing three diesel powered generating sets with space for a fourth which was never installed.

Wheal Vlow, Perranporth, *c.*1928. A group of miners taken near the portal of the mine adit. The figure in the front row second from the right is N.F.C. Burley the grandfather of Joyce Tregonning who kindly provided the photograph.

Wheal Droskyn, Perranporth, 1890s. Old men's workings in the cliffs at the western beach. These were part of Wheal Droskyn, which was an old and extensively worked mine. Note the many children posed in the photograph.

Wheal Droskyn, c.1900. Note the headgear on one of the shafts of the mine and Droskyn Castle in the top left of the photograph. Offshore can be seen Bawden Rocks otherwise known as Man and his Man – but always referred to locally as the Cow and Calf !

Perran Coombe stamps, Perranporth, 1924. A horse-drawn cart tipping a load of tin stone into the ore bin of the waterwheel driven stamps. Note the typical dress of the driver some seventy-eight years ago. (Photo H.G. Ordish)

Perran Coombe stamps, 1924. Viewed from below the road where the previous photograph was taken. The waterwheel driving the stamps is on the right. Water is also taken from the same leat via a launder to drive the small waterwheel which powers the rotating arms of the buddle on the left. The crushed ore from the stamps is fed on to the centre of the buddle by the 'V' shaped launder. This little stamps is typical of hundreds throughout the county which served as custom mills. They would accept parcels of ore from all and sundry and produce a tin concentrate for shipment to the smelters. (Photo H.G. Ordish)

Wheal Leisure, Perranporth, c.1900. The considerable waste burrows of the mine and the derelict engine house are all that remained at this time. In the foreground is a large parade of the Boys Brigade.

Wheal Leisure, Perranporth, March 1937. The hanging wall of this shaft collapsed and carried away the sollar and fill thus creating the vast cone to be seen here. The responsibility for making good this crater lay with the Mineral Lord i.e. the owner of the mineral rights and it is recorded that some 3,000 tons of sand was required for the operation.

Wheal Leisure, Perranporth, March 1937. A further view of the collapse of the shaft and it is rather interesting from a mining point of view to see the layering in the crater. From the top layer of mine dump material there followed beach sand, sub soil and finally killas rock. In the shaft sides there were drill holes which dated to the time when the shaft was being sunk. The building on the right is the back of Sully's Hotel.

Wheal Leisure, Perranporth, early 1960s. The Count House of this once extensive and productive copper mine. This photograph was taken only a short time before the building was demolished.

East Wheal Rose, near Newlyn East, 1920s. Actually this is North Rose but was worked as a part of the much larger East Wheal Rose. This house contained a pumping engine with a cylinder of 100in diameter which commenced working in 1884. It is worth mentioning the career of this enormous piece of machinery, which was built by Harvey & Co. of Hayle for Great Wheal Vor and put to work in 1854. In 1865 it was sold to the Hendre Lead Mines in Flintshire. After service here the East Wheal Rose Co. purchased the engine in 1881 but due to various delays it did not commence work until 1884. In 1888 the engine left Cornwall yet again to be used by the Cumberland Iron Mining & Smelting Co. As far as East Wheal Rose is concerned this was the last re-working of one of Cornwall's most famous lead and silver mines which, as it turned out, was not a success. In recent times most of the mine burrows have been landscaped and this is now the site of the narrow gauge Lappa Valley Railway leisure area. Happily the engine house and stack are still in existence.

East Wheal Rose, near Newlyn East, 1920s. Another view of the engine house and separate stack. This shows the unusual architecture of the building. The arrow slit windows were to the design of an architect who had come to live in Cornwall for health reasons. The new house for a 90in engine on Penrose's shaft at this time was also designed by him and had similar windows. When Penrose's engine ceased work it was bought back by her builders, Harvey & Co. of Hayle, who requested that the engine stay on the mine until they had obtained a purchaser. Two years later in 1888 Wheal Agar at Pool, Redruth, became the new owners and it was re-erected on their Robartes shaft. No doubt with economy in mind they also bought the metal window frames and built their new engine house with arrow slit windows.

West Chiverton Mine, near Perranzabuloe, c.1880. Showing the Count House and in the background the stamps engine and house. This engine had ceased working at this time as the mine was only operating on a limited scale and closed within a year or so.

West Chiverton Mine, near Perranzabuloe, c.1927. A view of the stamps engine house and boiler house at this one time prolific silver/lead mine. This house was wantonly destroyed as a military demolition exercise just after the end of the Second World War. The 26in cylinder rotary stamps engine was sold to the South Condurrow mine at Troon, near Camborne in 1881 and converted to a winding engine for hoisting on Marshall's shaft. In the background is the derelict engine house on Batter's shaft which had contained an 80in pumping engine. (Photo H.G. Ordish)

Silverwell Lead Mine (otherwise Wheal Treasure), near Mithian, 1935. The horizontal single cylinder engine was built by Bartle's Foundry and performed both winding and pumping duties. This little prospect was active in the first decade of the twentieth century but had closed by 1914. The plant lay idle for many years and was finally scrapped during the Second World War. Another photograph appears in Volume 2 of this series. Fortunately around the time this photograph was taken the late Bill Newton measured the engine and commenced to make a model from his drawings. It was not until forty years later that he and the author completed the model which is now in the Redruth Museum. Note Mithian Church on the skyline right. (Photo H.G. Ordish)

Lambriggan Mine, Perranzabuloe, 1927. A photograph taken in the early stages of the re-opening of the mine. Main shaft is surmounted by a windlass and a temporary canopy. In the background is a steam winch and a large portable engine with the flywheel and crankshaft not yet erected. This was supplied by Sykes & Co. of London who also provided this very neat and efficient plant for the un-watering of Wheal Coates at St Agnes earlier in the century.

Lambriggan Mine, Perranzabuloe, 1927. The portable engine is being enclosed and the pump gearing is well on the way to completion with the balance box and sweep rod clearly shown. A temporary small headgear and tripod surmount the shaft.

Lambriggan Mine, Perranzabuloe, 1927. The permanent headgear is erected on Main shaft and the boiler of the portable engine is still providing steam for pumping and the winch. The first Lancashire boiler is being installed with the ore bin and tramway to the dump on the right.

Lambriggan Mine, Perranzabuloe, 1927. It will be noted that in the headgear there are two sheave wheels. The larger one is for hoisting and the smaller one for capstan work in connection with the pitwork in the pump compartment of the two compartment shaft. The large boilered portable engine was supplying all the steam for the various functions at this time.

Lambriggan Mine, Perranzabuloe, 1928. By this time the portable engine and Sykes plant had been superseded by a very small horizontal steam winding engine, said by the late Jack Trounson to have been the most diminutive first motion winder that he had ever seen! In addition there was a horizontal engine formerly at the Kingsdown Mine, Hewas Water, driving the Cornish pitwork and a further horizontal engine driving the compressor. Steam for these engines was now provided by adequate permanent boiler capacity. On the extreme left of the scene are new foundations for a steam capstan which had previously been on the Engine shaft of East Pool Mine, near Redruth. To serve this new capstan a one-legged shears was built with the other end of the cap piece resting on the main frame of the headgear. Concurrent with this work the two compartment Main shaft was stripped down on the south side to provide a further hoisting compartment and more economical balanced winding.

Lambriggan Mine, Perranzabuloe, 1930. All the major operations in this re-working were based on Main shaft, which was deepened to 400ft. This view of the plant was taken just after the mine had closed and illustrates the headgear and capstan shears with the two sheave wheels for hoisting as described in the previous photograph. R.C.M. Robinson & Co. who undertook this venture did not achieve success for a variety of reasons the primary one being the prevailing low prices on the Metals market which as we now know heralded the great depression of the 1930s. An interesting footnote is the fact that in the 1970s an exploration company, a subsidiary of South Crofty, was considering the possibilities at Lambriggan.

Above: Wheal Ramoth, Perranporth, 1920s. The engine house shown at this very old mine formerly contained a 60in pumping engine which was erected around 1830 during a re-working of the property at that period.

Left: Wheal Ramoth, Perranporth, 1920s. A view from the front of the engine house showing the bob wall and an unusually large plug doorway. The mine was receiving some attention in the very early twentieth century but no work ensued. The house was demolished around 1939.

Opposite: Wheal Mexico, near Rejerrah, Perranzabuloe, 1920s. This was a small silver/lead mine which had an engine house close to the Treamble branch line. The ruins of the house and stack together with the track of the railway are seen here.

Treamble Mine, 1920s. Two miles NE of Perranporth. Although situated in an area which principally mined iron ore, a further product at Treamble was the soft talcky substance called Fullers Earth. Some 500 tons a year was sold for such diverse uses as refining of vegetable oils, a filler for paint, rubber and bituminous products. This scene shows the quarry terminal of an aerial ropeway built between the quarry and the mill where the material was refined.

Treamble Mine, 1920s. On the left is the Mill which treated the material to produce Fullers Earth. One of the aerial ropeway towers is visible in this photograph. This was about 1,600ft long and led to the quarry terminal seen in the previous photograph. The excavation in the background was known as Western Quarry.

Treamble Mine, 1937. Between 1859 and 1892 the mine worked intermittently and produced, under various companies, about 16,000 tons of brown haematite (iron ore), some lead ore together with small amounts of silver and zinc. The principal ore body is the Perran Iron lode This photograph shows the commencement of operations by Lloyds Perran Iron Co. who planned to mine the deposit by open-cast methods. A caterpillar tractor and scraper is at work. In the background is a portable air compressor.

Treamble Mine, 1937. An early type of bulldozer at work on the site in the period when Lloyds Perran Iron Co. was developing the mine.

Above: Treamble Mine, 1937. A further view of a tractor and scraper at work. The mine was served by a branch line nearly three miles long. This emanated from a junction at Shepherds on the Chacewater-Newquay branch railway but pre-dates that branch by many years. It originally formed part of the Cornwall Minerals Railway and was constructed in 1874.

Right: Treamble Mine, 1937. A photograph of the big pit and heavy earth moving machinery employed at that time with a view to re-working this iron ore deposit.

Below: Treamble Mine, c.1937. A panoramic view which illustrates the scope of the operations. On the left hand edge of the photograph is one of the engine houses on Great Retallack mine.

Treamble Mine, late 1930s. Illustrating the heavy earth moving equipment employed by Lloyds Perran Iron Co. when they were working the iron deposit. The anticipated lode widths were not met with after the excavation of a pit 250yds by 150yds and 7yds in depth. In 1940/1941 this company gave up the lease after open-working had been abandoned and underground mining commenced albeit in a small way. Thereafter the Government Home Ores Dept took up the property as by this time the Second World War had broken out and a shortage of iron ore was envisaged.

Treamble Mine, c.1938. In order to transport the ore to the Mill the company purchased two second-hand narrow gauge locomotives. These were 0-4-2 tank engines built by Kerr Stuart in 1903 and 1911 for Dolcoath Mine, Camborne, to work their extensive tramway system. Since 1921 they had lain idle in their engine shed and consequently were in good condition, not having been exposed to the elements. We see one of the locomotives being re-railed by a traction engine at Treamble.

Treamble Mine, c.1940. Viewed from the west and showing the big pit. In this photograph can be seen the beginning of underground operations as continued by the Government Home Ores Dept in the Second World War. All work on the site finally ceased about 1943.

Phoenix Mine, near the northern end of Perranporth beach. This photograph is believed to have been taken in the 1920s and shows the pumping engine house. As the mine closed in 1874 the roof, although having lost most of its slates, is in remarkably good condition.

Wheal Golden, near Crantock c.1900. This was a most decorative engine house which once contained a 50in pumping engine and was destroyed by the military during the Second World War. Close inspection will reveal two other engine houses in this scene. It is clearly a summer day judging by the dress of the lady and gentleman.

Mount Mine (also known as Perran Iron Mine), near Cubert. Taken in the summer of 1916 by the adit portal with a group of ladies sitting on 2ft gauge mine cars. Each one is holding a miner's tallow candle but it is highly unlikely that they actually ventured underground especially in view of their attire! The short-lived re-opening of the mine at this time was not a success. In the nineteenth century however the mine had been a moderate producer of iron ore. It will be noted that the tunnel portal is unusually large. The reason for this is that in an earlier working this adit was driven large enough to accommodate standard gauge (4ft 8½in) mineral waggons. This track then connected with the Treamble branch line thus saving trans-shipment of ore. (Photo courtesy of the late Tony Barrett)

Gravel Hill Mine, 1937. Situated at the northern end of Perran beach. This was one of the many mines developed on the Perran Iron Lode which outcrops in the cliffs near this scene. From here the lode has been traced eastwards for approximately four miles. Among the captions for the Treamble mine photos contained in this volume reference is made to the railway branch from Shepherds junction to Treamble. It may be of interest to mention that a further extension from Treamble to Gravel Hill was built in 1874. This was constructed with the consent of the landowners but without Parliamentary Powers! The line of just over a mile in length remained in use until 1888 when the track was removed. (Photo courtesy Royal Institution of Cornwall)